Introducing Algebra 1:
Number Patterns and Sequences

Dr Graham Lawler MA. PhD, Adv. Dip Educ., Cert Ed CATL.

Mr. Educator
Making Learning Fun!

978-1-84285075-6

First Published by Mr Educator Books, PO Box 225, Abergele Conwy County LL18 9AY

Mr Educator is a division of GLMP Ltd

Computer software screenshots courtesy of the Microsoft Corporation

Some photos courtesy of free digital photos: Timeless photography, DJ Crodin, Tom Clare

Mr. Educator
Making Learning Fun!

Teacher Notes

As a child I was completely baffled by the manner in which I was introduced to algebra, it simply did not make sense. As a teacher I can now wince at the errors made by the teacher as he struggled to make children understand.

The whole emphasis here is to use the concept of assessment for learning and to move children from where they are to where they are capable of being (see Vygotsky's Zone of Proximal development).

The aim of this series is to support teachers in developing mathematical thinking in children. To this end the activities are designed to move children from concrete operational stages of thinking. The emphasis throughout is to move to subsequent short handing.

Teachers are advised to engage children/students in meaningful discussion. We have used a paradigm of 'convince yourself then convince another person' throughout the series. In trials this worked well BUT it can be a source of frustration to those who do not have the language to convey their findings. Therefore it may be a requirement for sensitivity and a pre-cursor to engaging in this activity is to state that if the other person does not understand, the explanation is not good enough or the mathematics is wrong. In trials some parents complained that their offspring were casting aspersions at the parents' intellectual capabilities when the parents confessed to not understanding and at least one family heard the refrain 'Granny you're thick!' hardly beneficial to good family relationships.
We therefore urge all colleagues to make a point of teaching the children how to communicate effectively as an ongoing process throughout the series.

The final point to emphasise is that, contrary to many children's' expectations this is a fun activity and we believe that your class will benefit if you re-enforce the fact that it is fun. Parents also need to be aware of what this means for the children and we therefore recommend that parents evenings/afternoons etc are used as an opportunity to inform parents of the developments in this part of the mathematics curriculum. The function of the quiz is to act as a diagnostic tool, make sure students are aware that the result is not important but that they must get the Check correct.

Most of all I want you and your charges to enjoy this experience, and then the work in this book will be worthwhile.

Regards

Graham Lawler

Quiz

1. Which times tables are these answers in?

a 12, 14, 16, 18   two times   three times   five times   seven times

b 3, 6, 9, 12   two times   three times   five times   seven times

c 5,10,15,20,25   two times   three times   five times   seven times

d 7,14,21,28   two times   three times   five times   seven times

2. Circle the even numbers in this list

   1   2   3   4   5   6   7   8

3. Circle the odd numbers in this list

   1   2   3   4   5   6   7   8   9

4. What are the odd numbers that are bigger than 12 but smaller than 20?

5. Which numbers greater than 100 but less than 110 are even numbers?

6. What is the next odd number after 99?

7. What is the even number before 101?

8. What two odd numbers added together make 10?

9. What two even numbers added together make 12, when one number is bigger than the other?

10. Are there two even but different numbers that when added together make 8?

Starting Points: You need to be able to calculate with negative numbers to be successful with the mathematics in this book.

Patterns in Tables

There are some brilliant patterns in times-tables.
The purpose of this section is to get you thinking about tables and looking for the 'sameness' in the tables.
Don't worry you will see it and it is fun!

1	2	3	4	5	6	7	8	9	10
11	12	13	14	15	16	17	18	19	20
21	22	23	24	25	26	27	28	29	30
31	32	33	34	35	36	37	38	39	40
41	42	43	44	45	46	47	48	49	50
51	52	53	54	55	56	57	58	59	60
61	62	63	64	65	66	67	68	69	70
71	72	73	74	75	76	77	78	79	80
81	82	83	84	85	86	87	88	89	90
91	92	93	94	95	96	97	98	99	100

1. On the table above shade in all of the answers in the two times table.

2. Complete this sentence:

The answers I have shaded for question 1 are called the

MULTIPLES OF 2 AND THEY END IN

So now you know a new word, ' Multiple'. It is almost the same as the word 'multiply' and that is a clue. Multiples are the answers in the times table, so the multiples of 2 are the answers in the two times table, the multiples of 3 are the answers in the three times table and so on.

1	2	3	4	5	6	7	8	9	10
11	12	13	14	15	16	17	18	19	20
21	22	23	24	25	26	27	28	29	30
31	32	33	34	35	36	37	38	39	40
41	42	43	44	45	46	47	48	49	50
51	52	53	54	55	56	57	58	59	60
61	62	63	64	65	66	67	68	69	70
71	72	73	74	75	76	77	78	79	80
81	82	83	84	85	86	87	88	89	90
91	92	93	94	95	96	97	98	99	100

3. On this grid, shade in the multiples of the 3 times table

Complete this sentence:

The answers I have shaded for question 1 are called the

MULTIPLES OF 3 AND THEY END IN

Making Learning Fun!

So now you know the multiples of 2 and 3 up to 100, but what is the same about them?

Let's look at the multiples of 2 first.

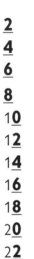

2
4
6
8
1**0**
12
14
16
18
2**0**
22
2**4**
2**6**
2**8**

Look at the underlined bolded numbers in this list of multiples of 2. Write down what you notice.

4. I notice that these numbers form a pattern of _____

5. This means the next multiple of 2 after 28 must be _____

6. Write down the next multiple of 2 after 6 _____

7. What is the multiple of 2 before 9? _____

8. What are the multiples of 2 between 15 and 25? _____

9. Is 50 a multiple of 2? _____

10. Is 51 a multiple of 2? _____

OK so now you know that multiples of 2 always end in 0, 2, 4, 6 or 8
So you can always tell if a number is a multiple of 2, just look to see what it ends in.
Remember multiples of 2 are also called EVEN NUMBERS!

11. Circle the numbers that are multiples of 2

1   3   5   7

2   99   12   14

49   22   37

15   18   29   16   44

102   199   300   501

77   89   109   50   66

12. What is the other name for multiples of 2? _____

Mr. Educator
Making Learning Fun!

Now we will look at the multiples of 3, remember these are the answers in the three times table.

Look at the row of numbers here.

3   6   9   12   15   18   21   24   27   30   33   3

1. Write down the next three multiples of 3.

---

2. Now look at this, add the digits, what do you notice?

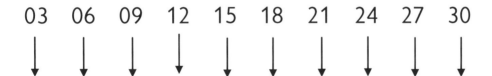

03   06   09   12   15   18   21   24   27   30

↓   ↓   ↓   ↓   ↓   ↓   ↓   ↓   ↓   ↓

0+3 =

0+6 =

0+9 =

1+2 =

1+5 =

1+8 =

2 +1=

2+4 =

2+7 =

3+0 =

What do you notice?
When I add the digits in the multiples of 3, I get pattern of _____, _____, _____

So you now have a really *cool* way of knowing
if a number is a multiple of 3.
Just add up the digits and if they come
to 3, 6 or 9 then it must be a multiple of 3.

3.   By adding the digits together, work out which of these numbers are multiples of 3.

15   22   33   65   89   99   12

103   46   78   36   34   38

44   56   59   21   105   18

Use this space for your working.

4. Look at your answers for question 3, are all multiples of 3 odd numbers?
First convince yourself , then convince another person.
Write down a sentence that explains your answer.

_____

5.

Ok so we have looked at multiples of 2 and 3, now we need to look at more numbers. In this grid below, shade in all of the multiples of 4.

1	2	3	4	5	6	7	8	9	10
11	12	13	14	15	16	17	18	19	20
21	22	23	24	25	26	27	28	29	30
31	32	33	34	35	36	37	38	39	40
41	42	43	44	45	46	47	48	49	50
51	52	53	54	55	56	57	58	59	60
61	62	63	64	65	66	67	68	69	70
71	72	73	74	75	76	77	78	79	80
81	82	83	84	85	86	87	88	89	90
91	92	93	94	95	96	97	98	99	100

Write down a sentence that explains the pattern you are seeing in the grid.

_____

6. What are the multiples of 4 between 50 and 60?

_____

7. What is the next multiple of 4 after 100?

_____

8.

1	2	3	4	5	6	7	8	9	10
11	12	13	14	15	16	17	18	19	20
21	22	23	24	25	26	27	28	29	30
31	32	33	34	35	36	37	38	39	40
41	42	43	44	45	46	47	48	49	50
51	52	53	54	55	56	57	58	59	60
61	62	63	64	65	66	67	68	69	70
71	72	73	74	75	76	77	78	79	80
81	82	83	84	85	86	87	88	89	90
91	92	93	94	95	96	97	98	99	100

On this grid shade in all of the multiples of 5.
Write a sentence that describes the pattern you see.

_____

9.

1	2	3	4	5	6	7	8	9	10
11	12	13	14	15	16	17	18	19	20
21	22	23	24	25	26	27	28	29	30
31	32	33	34	35	36	37	38	39	40
41	42	43	44	45	46	47	48	49	50
51	52	53	54	55	56	57	58	59	60
61	62	63	64	65	66	67	68	69	70
71	72	73	74	75	76	77	78	79	80
81	82	83	84	85	86	87	88	89	90
91	92	93	94	95	96	97	98	99	100

On this grid shade in all of the multiples of 7. Write a sentence that describes
the pattern you see.

_____

10. Follow these instructions
    a)   add the 2 digits together
    b)   if you get a two –digit or more answer, add again until you get a single digit answer
    c)   What pattern can you find?

02	04	06	08	10	12
14	16	18	20	22	24
26	28	30	32	34	36
38	40	42	44	46	48
50	52	54	56	58	60
62	64	66	68	70	72
74	76	78	80	82	84
86	88	90	92	94	96
98	100				

My pattern is:

_____

_____

_____

_____

11. Look at other times tables, what patterns are there in the 5 times table, the 6 times table, the 9 times table. What about other times tables?

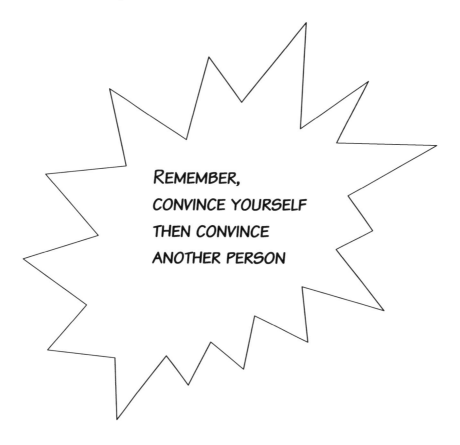

REMEMBER,
CONVINCE YOURSELF
THEN CONVINCE
ANOTHER PERSON

12.

a) This section of a spreadsheet shows some results from a table patterns investigation.
i) Which table has been investigated?
ii) What results does this spreadsheet tell us?
iii) Can you use a spreadsheet in your investigation?

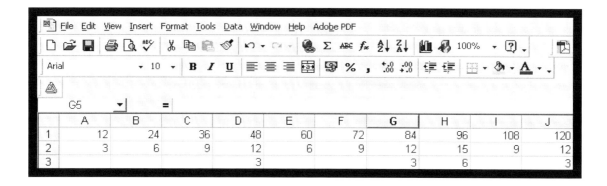

	A	B	C	D	E	F	G	H	I	J
1	12	24	36	48	60	72	84	96	108	120
2	3	6	9	12	6	9	12	15	9	12
3				3			3	6		3

b) Prepare a talk for your class, using Powerpoint or a similar program

c) Write a report on your findings.

Teacher Notes

On the previous page, question 12 is an extension piece for the more able child.

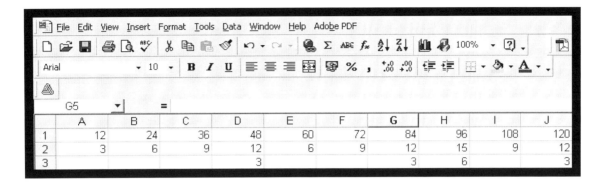

Here the spreadsheet is investigating the 12 times table. The cells are referenced by the column header and then the line reference. So 12 is in cell A1.

The digits of the first line have been added together to give the second line, so.

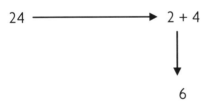

Where the first addition has resulted in a double-digit number, we have added again.

For instance

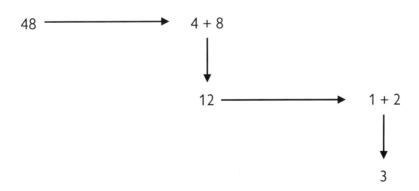

Invastigating number patterns using computers

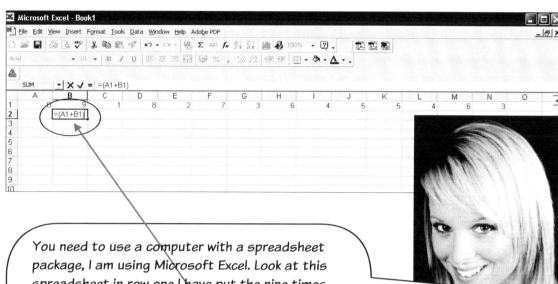

You need to use a computer with a spreadsheet package, I am using Microsoft Excel. Look at this spreadsheet in row one I have put the nine times table, as 09, 18, 27 and so on, with one number in each cell.

Now look at the formula in cell B2 . This is adding the contents of cell A1 to the contents of cell B1 and putting the answer here in B2.

Put the nine times table into the first row like I have done.

Copy the formula in B2 and then press enter, what happens? Can you explain why it happens?

Can you work out the formula to go into D2 to add together C1 and D1?

Try this formula, (remember all formulas in a spread sheet start with an = sign and then a bracket so it will start =(

Don't forget to close the bracket, should formula should look like =(Cell ref + cell ref)

Where else would you put formulas in the second line?

Have a go. Do they work?

Try another times table, say the 12 times table.

MR. Educator
Making Learning Fun!

Multiplying with Spreadsheets

Look at your keyboard, on the right hand side usually above the number block there is a button marked *.

This is how you write a multiplication sign, so 3 x 4 on a computer is written as 3*4

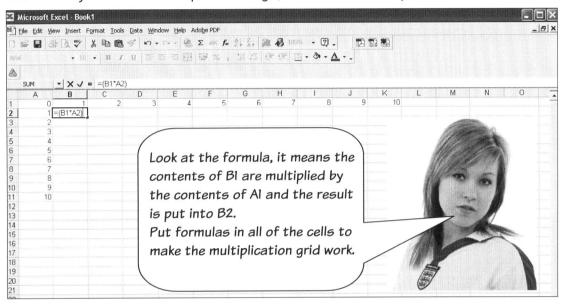

Look at the formula, it means the contents of B1 are multiplied by the contents of A1 and the result is put into B2.
Put formulas in all of the cells to make the multiplication grid work.

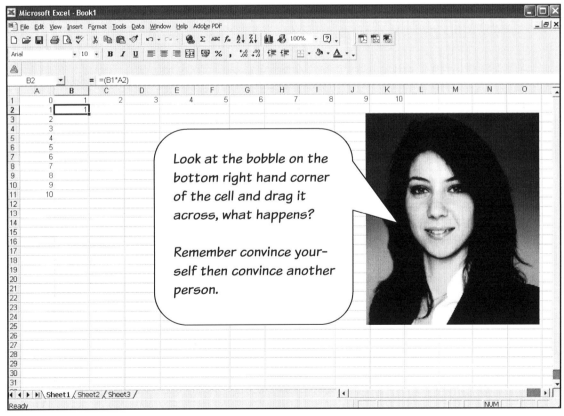

Look at the bobble on the bottom right hand corner of the cell and drag it across, what happens?

Remember convince your-self then convince another person.

Write a report on your findings.

Check

The purpose of a check is to for you to see how well you are doing.

So try answering these questions

1. Write down the multiples of 2 between 15 and 21

_____

2. Is 120 a multiple of 3?

_____

3. What are the multiples of 2 less than 20 that are also multiples of 3?

_____

4. Find the multiple of 9 that is bigger than 18 but less than 50, which is also a multiple of 2.

_____

5. Are there any numbers less than 50 that are multiples of 7 and 8?

_____

Quiz   21

1.   Write down the multiples of 2 between 15 and 21

_____

2.   Is 120 a multiple of 3?

_____

3.   What are the multiples of 2 less than 20 that are also multiples of 3?

_____

4.   Find the multiple of 9 that is bigger than 18 but less than 50, which is also
      a multiple of 2.

_____

5.   Are there any numbers less than 50 that are multiples of 7 and 8?

_____

Fascinating Number Nine

1.

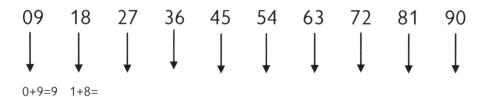

Look at the 9 times table, add up the digits; then write down what you notice.

| 09 | 18 | 27 | 36 | 45 | 54 | 63 | 72 | 81 | 90 |

0+9=9   1+8=

Write down what you notice.

2.

Look at the arrangement for the 9 times table below, work down the left hand column and then up the right hand side. What do you notice?

09	54
18	63
27	72
36	81
45	90

I notice that there is a pattern and the pattern is....

_____
_____
_____
_____
_____
_____
_____
_____

Multiples and Factors

So now we know that a multiple is the answer in a times table, so the multiples of 2 are the answers in the 2-times table and the answers in the 3-times table are multiples of 3, the answers in 4-times table are multiples of 4 and so on.

So now we need a new word; that is FACTOR. A factor is a number that goes into another number with no remainder, so 3 is factor of 6 because 3 goes into 6 with no remainder.

So all of the factors of 6 are 1,2,3 and 6 because these are all of the numbers that go exactly into 6 with no remainder.

Write down the factors for each of these numbers, the first is done for you.

1.   8   ⟶   1, 2, 4, 8

2.   10   ⟶   _____

3.   20   ⟶   _____

4.   30   ⟶   _____

5.   18   ⟶   _____

Compare your answers with another person; make sure that you have all of the factors of each of the numbers.

Factor Diagrams

Did you notice that each number is a factor of itself?
Look back at the answers you gave in the last question. Each number is a factor of itself; make sure you understand this point. Remember convince yourself, then convince another person.

1. ⟶ **3.**

6.

This is a factor diagram; the lines show which numbers are factors of other numbers.
Notice the loops: this means that each number is a factor of itself. Look at how we make it.

I is a factor of 3, so we draw an arrow between I and 3.

1 ⟶ 3

Remember each number is a factor of itself so each number should have a loop pointing back to itself.

1. ⟶ **3.**

Mr. Educator
Making Learning Fun!

Finally, we draw arrows between 1 and 6 because 1 is a factor of 6 and an arrow is drawn between 3 and 6 because 3 is a factor of 6.

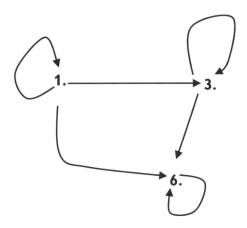

So this is the final factor diagram. Look at the arrows, they show you which numbers are the factors and what they are factors of, so 1 has arrows to itself, 3 and 6, so 1 is a factor of 1,3 and 6.

Complete these factor diagrams.

1.

    1   4

        8   12

2.

    14   6   16

    18   24   25

3.

2   8   9

5   16   18

4.

5

15   3

25

30   36   2

5.

1   6   8   10

16   25

49   7   85   26

14   28   32

In this space make up a factor diagram of your own and challenge another person to complete it. At the same time you can complete their factor diagram.

19	18	17	16	15	14	13	12	11	10
20	51	50	49	48	47	46	45	44	9
21	52	75	74	73	72	71	70	43	8
22	53	76	91	90	89	88	69	42	7
23	54	77	92	99	98	87	68	41	6
24	55	78	93	100	97	86	67	40	5
25	56	79	94	95	96	85	66	39	4
26	57	80	81	82	83	84	65	38	3
27	58	59	60	61	62	63	64	37	2
28	29	30	31	32	33	34	35	36	1

You need two copies of this grid.

1. Shade in all of the multiples of 5 on the first grid. Which line has no multiples of 5 in it? _____

2. On the second grid shade in all of the multiples of 9. Which lines have no multiples of 9 in it?_____

Special Sequences

Look at this pattern.

 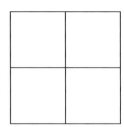 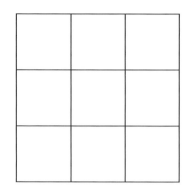

1.   4   9

Look at the pattern, in the space below, draw the next three in the sequence, what do you notice about the shapes that are formed in this sequence.

Remember, Convince yourself, then convince another person

So did you find the pattern?

 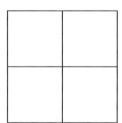 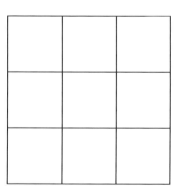

Look at the pattern here, you have a 1 x 1= 1, then a 2 x 2 = 4, then a 3 x 3 = 9, so the next one must be a 4 x 4 = 16, then 5 x 5 =25 and so on. They are all squares, look at the shape, in each diagram they are squares.

This is the sequence of square numbers.

Look again, without the diagrams

1 x 1 = 1 so 1 is a square number   2 x 2 = 4 so 4 is a square number

3 x 3 = 9 so 9 is a square number   4 x 4 = 16 so 16 is a square number

5 x 5 = 25 so 25 is a square number

complete the rest

6 x 6 =

7 x 7 =

8 x 8 =

9 x 9 =

10 x 10 =

11 x 11 =

12 x 12 =

13 x 13 =

14 x 14 =

15 x 15 =

Top tip: it is a good idea to learn the first 15 square numbers by heart, so that you recognise them whenever you see them in an assessment or exam.

Mr. Educator
Making Learning Fun!

Another Special Sequence

Look at this sequence.

 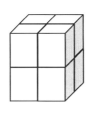 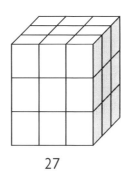

1   8   27

Look at the pattern; can you see how the pattern is made up? (hint look at the length, width and height of each cube).

Let's look at the sequence in another way

$1 \times 1 \times 1 = 1$

$2 \times 2 \times 2 = 8$ (notice this is $2 \times 2 = 4$ then $4 \times 2 = 8$)

$3 \times 3 \times 3 = 27$

now workout

$4 \times 4 \times 4 =$

$5 \times 5 \times 5 =$

$6 \times 6 \times 6 =$

What about $10 \times 10 \times 10$?

Don't be surprised if you found this one a bit tough, I did when I did it in my school. Look at the shape these numbers make. Guess what their name is, they are called CUBE numbers.

Special Numbers

Name _____   Teacher's Name _____

1	2	3	4	5	6	7	8	9	10
11	12	13	14	15	16	17	18	19	20
21	22	23	24	25	26	27	28	29	30
31	32	33	34	35	36	37	38	39	40
41	42	43	44	45	46	47	48	49	50
51	52	53	54	55	56	57	58	59	60
61	62	63	64	65	66	67	68	69	70
71	72	73	74	75	76	77	78	79	80
81	82	83	84	85	86	87	88	89	90
91	92	93	94	95	96	97	98	99	100

Follow these instructions on the grid.

1.   Put a circle around 1 and cross it out.

2.   Put a circle around 2 and leave it. Shade in all of the other multiples of 2.

3.   Put a circle around 3 and leave it. Shade in all of the other multiples of 3.

4.   Put a circle around 5 and leave it. Shade in all of the other multiples of 5.

5.   Put a circle around 7 and leave it. Shade in all of the other multiples of 7.

6.   Now circle all of the remaining numbers.

Now write out all of the numbers that are not coloured in here and work out the factors of that number, the first one is done for you.

Number	Factor
2	1,2
3	
5	
7	
11	

Look at the factors of these numbers, what do you notice that is the same about them? Remember, convince yourself then convince another person.

So now we know that all of these numbers have only 2 factors and the factors are different, how cool is that!

They are called PRIME NUMBERS

These numbers are very important in real life because they have two factors that are different; they form the basis for security on the Internet. Every time someone uses a secure site on the net, they are using combinations that can use prime numbers.

Prime numbers have fascinated mathematicians for hundreds of years and a famous mathematician called Pierre De Fermat really made people cross over his work on prime numbers.

He said he found a really nice way of working something out but there wasn't enough room to write it down. No one could work out what he meant until a man called Andrew Wiles solved it. Fermat wrote his note in about 1630 but it took until the 1990's for Andrew Wiles to solve it.

How many years is it between 1630 and 1995 when Professor Wiles worked out the answer to Fermat's maths?

Pierre De Fermat

Andrew Wiles

Triangle Numbers

Work out the next three numbers in this sequence.

What are these numbers called? (Look at the pattern they make)

Fibonacci

This is a statue of a man called Fibonacci. He was a very important mathematician in the history of mathematics and is very famous for a certain sequence.

He was born around AD 1170/ACE 1170 and when he grew up he travelled to the Mediterranean to study mathematics under the most famous Arab mathematicians of the time. He came back from his travels in around 1200 and in 1202 wrote and published a book called Liber Abaci (Book of Abacus or Book of Calculation). This introduced the numbers we use today, into Europe.

Look at the sequences below and work out the missing numbers

1.

1	1	2		5	8

2.

1	2			8	

3.

2		5	8		

4.

8					

More on Sequences

In Maths a sequence is a series of numbers that follows a rule, the trick is in finding the rule. In this section we are going to look at some sequences, I really like doing this in my school it is fun!

In these questions, find the next three numbers in the sequence and say what the rule is, oh and check out my cool dude hat!

1.   1   3   5   7   9   11   ___, ___, ___

The rule is _____

2.   2   4   6   8   10   ___, ___, ___

The rule is _____

3.   5   10   15   20   25   30   35   ___, ___, ___

The rule is _____

4.   7   14   21   28   35   42   ___, ___, ___

The rule is _____

5.   9   18   27   36   45   54   63   ___, ___, ___

The rule is _____

6.   12   24   36   48   60   72   84   ___, ___, ___

The rule is _____

7.   15, 30, 45, 60, 75, ___, ___, ___

The rule is _____

8.   16, 32, 48, 64, 80, ___, ___, ___

The rule is _____

9.   13, 26, 39, 52, 65, ___, ___, ___

The rule is _____

10.   8, 16, 24, 32, 40, ___, ___, ___

The rule is _____

11.   1, 2, 3, 4, 5, 6, 7, 8 ___, ___, ___

The rule is _____

12.   3, 6, 9, 12, 15, 18, ___, ___, ___

The rule is _____

13.   29, 25, 21, 17, 13, ___, ___, ___

The rule is _____

14.   49, 42, 35, 28, ___, ___, ___

The rule is _____

15.   41, 39, 37, 35, ___, ___, ___

The rule is _____

16.   80, 71, 62, 53, ___, ___, ___

The rule is _____

17. 100, 81, 64, 49, 36, __, __,___

The rule is _____

18. 01, 11, 21, 31, 41, _____, ___, _____

The rule is _____

19. 6, 66, 6666, 66 666, ____, ____, ____

The rule is _____

20. 0, 5, 0, 10, 0, 15, ____, ____, ____

The rule is _____

21. 6, 12,___, 24, 30, ___, 42,___

The rule is _____

22. 500, 450,___, 350, 300,____, 200, ____

The rule is _____

23. Can you make a sequence with five numbers, which has 6 and 12 in it?

_____

24. Can you make a sequence with five numbers that has 5 and 20 in it?

_____

25. Can you make a sequence with five numbers that has 9 and 16 in it?

_____

26. Can you make a sequence with five numbers that has a 6 and a 20 in it?

_____

27. Can you make a sequence with five numbers that has 50 and 101 in it?

_____

28. Can you make a sequence with five numbers that has 100 and 501 in it?

_____

29. Can you make a sequence with five numbers that has 30 and 90 in it?

_____

30. Can you make a sequence with five numbers that has 10 and 51 in it?

_____

More Sequences

						75		45	30	15

13	22	31		58						

				40	47					

		60	52	44						

2.5	4	5.5			10		13			

	14		35		56					

								36		12

			8							

				60						

Number sequences - counting in 2s

2	4	6			12				20	

3	5	7		11			17			

32	34	36			42		46			

31	33		37		41	43			49	

Number sequences - counting in 3s

3	6	9			18	21		27		

5	8		14	17	20			29		

11	14		20	23		29	32			

41	44		50							

0	6	12			30	36				60

7	14	21		35	42	49	56			

14	28		56	70	84					

10	17		31	38						

14	19	24			39					

36	34	32		28	26			20		

		-10	-13	-16		-22	-25		-31	

				21	24	27				

		46	50		58	62			74	

-94	-90	-86		-78	-74	

23		33		43		53		63		73

	16		24		32		40		48	

14	28		56	70	84					

93	85		69		53	45	37			

		43		44						

74		62		50						

100		201								

	25		50							

	22.5		47.5							

	150		194							

Word Search

X	F	D	X	C	F	V	B	N	A	T	U	O	K	L	R	D	V	S	C
A	W	N	Y	J	G	L	D	I	B	F	G	H	M	N	O	A	D	D	G
M	U	A	Z	T	U	J	V	M	N	K	L	P	E	R	C	E	N	T	I
U	B	T	G	B	V	J	R	C	A	K	D	D	G	J	K	R	V	F	D
L	M	A	F	C	I	L	J	K	Y	A	S	F	F	W	H	S	A	G	K
T	A	B	C	A	L	B	U	I	O	F	R	G	T	W	C	H	J	S	A
I	R	S	T	Y	C	V	F	G	H	J	K	L	D	E	F	R	G	Y	M
P	R	H	Q	W	F	T	B	Q	S	D	E	F	E	Y	R	B	Y	E	F
L	F	J	A	S	D	A	O	U	Y	R	F	H	E	F	W	J	K	P	K
I	E	E	P	D	V	N	M	R	D	J	S	E	D	I	V	I	D	E	Y
C	C	D	A	G	S	E	Z	X	C	M	P	M	S	D	W	G	X	R	A
T	W	A	T	H	C	S	Z	X	V	N	R	W	D	H	V	S	W	C	W
I	X	V	S	Y	W	A	T	Y	U	I	E	G	H	J	W	R	A	N	D
O	Z	S	U	B	T	R	A	C	T	Z	A	F	G	H	J	J	K	T	L
N	C	K	R	Y	J	Q	B	N	D	V	D	S	F	H	D	D	W	A	G
J	Z	I	T	U	U	U	W	T	Y	U	H	L	I	E	C	T	Z	G	E
L	Z	O	G	M	U	L	T	I	P	L	E	B	N	M	D	U	I	E	D
K	C	P	H	O	Z	O	I	D	F	H	E	R	G	W	E	G	H	S	C
R	B	S	J	Q	X	P	G	H	Y	E	T	I	S	A	V	N	D	G	HJ
W	N	D	E	A	C	R	M	K	L	E	F	T	G	V	W	S	D	G	A

Cross Number

1		2			3			4	5		6
					7						
8							9				
					10	11					
				12	13			14			
		15									
16				17	18					19	
		20	21					22			
23						24					
		25									
26					27						

Across

1 one hundred and fifty five thousand five hundred and fifty five in numbers

2 5001 + 554

3 2 x 2 then add 1

4 1000 + 101

5 110 -9

7 1000 + 250

8 1/2 a million in figures

9 one thousand nine hundred and ninety nine in figures

10 four thousand five hundred in figures

12 66 x 10

13 5 x 12

16 220 -18

17 sixty five thousand in figures

18 5 x 1000

20 666-222

22 4 x 11

23 20 000 + 5552

24 fifteen thousand in figures

25 33 x 20

26 2900 -1 6

27 one hundred and fifty thousand in figures

Down

1 twelve thousand and one + four hundred and ninety nine

2 five thousand + one hundred and one

3 fifty one thousand and forty six in figures

4 1600 - 90

5 one million and ninety thousand five hundred in figures

6 2000 - 10

7 one thousand and fifty - 4 in figures

8 50 x 10

10 23 x 2

11 5 x 10

12 six million five hundred and sixty thousand four hundred and twenty in figures

14 50 x 10

15 two million two hundred and thirty four thousand five hundred and sixty eight in figures

16 2 x 11

19 99 999 + 1

20 5000 - 432

21 4500 + 64

22 1999+1

23 111 x 2

25 (5 x 12) + 8

Answers

P6 Quiz

1a two times **1b** three times **1c** five times **1d** seven times

2. 2,4,6,8 **3.** 1,3,5,7,9 **4.** 13,15,17,19

5. 102, 104, 106, 108 **6.** 101 **7.** 100

8. 7 + 3 (or 5 + 5 hmm!) **9.** 8 + 4 **10.** yes 6+2

P7

1.

1	2	3	4	5	6	7	8	9	10
11	12	13	14	15	16	17	18	19	20
21	22	23	24	25	26	27	28	29	30
31	32	33	34	35	36	37	38	39	40
41	42	43	44	45	46	47	48	49	50
51	52	53	54	55	56	57	58	59	60
61	62	63	64	65	66	67	68	69	70
71	72	73	74	75	76	77	78	79	80
81	82	83	84	85	86	87	88	89	90
91	92	93	94	95	96	97	98	99	100

2. 0, 2, 4, 6, 8

P8

3.

1	2	3	4	5	6	7	8	9	10
11	12	13	14	15	16	17	18	19	20
21	22	23	24	25	26	27	28	29	30
31	32	33	34	35	36	37	38	39	40
41	42	43	44	45	46	47	48	49	50
51	52	53	54	55	56	57	58	59	60
61	62	63	64	65	66	67	68	69	70
71	72	73	74	75	76	77	78	79	80
81	82	83	84	85	86	87	88	89	90
91	92	93	94	95	96	97	98	99	100

1, 2, 3, 4, 5, 6, 7, 8, 9

P9

4. 2,4,6,8,0 **5.** 30 **6.** 8

7. 8 **8.** 16, 18, 20, 22, 24

9. yes (is 50 mult if 2) **10.** No (is 51 mult of 2)

P10

11.

1   3   5   7

②   99   ⑫   ⑭

49   ㉒   37

15   18   29   ⑯   ㊽

⑩⑫   199   ③⓪⓪   501

77   89   109   ㊿   ⑥⑥

12. Even numbers

P11

1. 39, 42, 45   **2.** pattern is 3,6,9

3. **15**   22   **33**   65   89   **99**   **12**
103   46   **78**   **36**   34
44   56   59   **21**   **105**   **18**

4. No they are not all odd numbers, 24 is 3 x 8 and is an even.

5.

1	2	3	**4**	5	6	7	**8**	9	10
11	**12**	13	14	15	**16**	17	18	19	**20**
21	22	23	**24**	25	26	27	**28**	29	30
31	**32**	33	34	35	**36**	37	38	39	**40**
41	42	43	**44**	45	46	47	**48**	49	50
51	**52**	53	54	55	**56**	57	58	59	**60**
61	62	63	**64**	65	66	67	**68**	69	70
71	**72**	73	74	75	**76**	77	78	79	**80**
81	82	83	**84**	85	86	87	**88**	89	90
91	**92**	93	94	95	**96**	97	98	99	**100**

Vertical lines with gaps.

6. 52, 56   **7.** 104

8. Vertical lines

1	2	3	4	**5**	6	7	8	9	**10**
11	12	13	14	**15**	16	17	18	19	**20**
21	22	23	24	**25**	26	27	28	29	**30**

31	32	33	34	35	36	37	38	39	40
41	42	43	44	45	46	47	48	49	50
51	52	53	54	55	56	57	58	59	60
61	62	63	64	65	66	67	68	69	70
71	72	73	74	75	76	77	78	79	80
81	82	83	84	85	86	87	88	89	90
91	92	93	94	95	96	97	98	99	100

9.

1	2	3	4	5	6	7	8	9	10
11	12	13	14	15	16	17	18	19	20
21	22	23	24	25	26	27	28	29	30
31	32	33	34	35	36	37	38	39	40
41	42	43	44	45	46	47	48	49	50
51	52	53	54	55	56	57	58	59	60
61	62	63	64	65	66	67	68	69	70
71	72	73	74	75	76	77	78	79	80
81	82	83	84	85	86	87	88	89	90
91	92	93	94	95	96	97	98	99	100

Diagonal lines

P15   **10.** 2,4,6,8,1,3,5,7,9   **P16**   **11.** Student-by-student

12.   **i,** 12's   **12. ii,** pattern of 3, 6, 9

P18   Adds the contents of a1 and b1
=(C1 + D1)
Where else would you put formulas? C2, D2 etc.

P19   Students should create a multiplication grid.
Dragging the bobble will replicate and update the formula with a new cell ref and therefore complete the calculations.

P20   **Check**
   1.   16, 18, 20   **2.**   yes   **3.**   6, 12, 18
   4.   36   **5.**   no

P21   **Quiz**
   1.   1,2,3,6   **2.**   1,2,4,5   **3.**   63
   4.   20   **5.**   4 and 9

P22
   1.   +9   **2.**   'mirror image'

P23

1.   1,2,5,10   **2.**   1,2,4,5,10,20   **3.**   1,3,5,6,10,15,30

4.   1,2,3,6,9,18

P25
1.

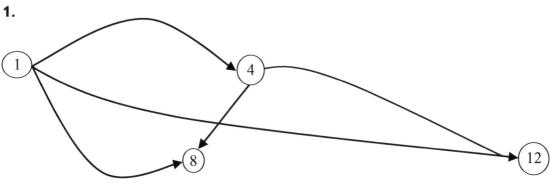

2.

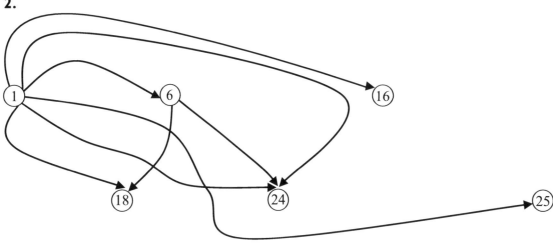

3.

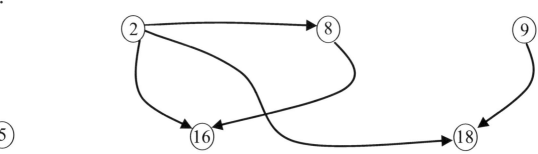

4.

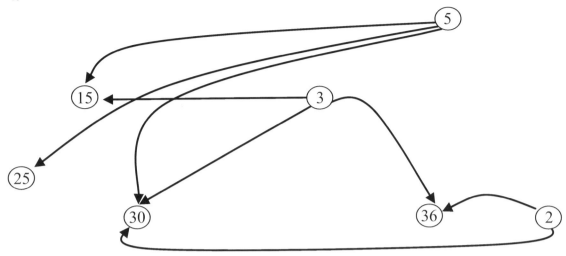

5.

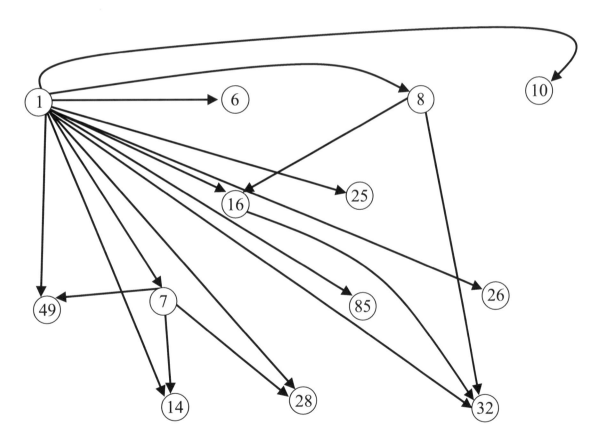

P28   Shading in fives

19	18	17	16	15	14	13	12	11	10
20	51	50	49	48	47	46	45	44	9
21	52	75	74	73	72	71	70	43	8
22	53	76	91	90	89	88	69	42	7
23	54	77	92	99	98	87	68	41	6
24	55	78	93	100	97	86	67	40	5
25	56	79	94	95	96	85	66	39	4
26	57	80	81	82	83	84	65	38	3
27	58	59	60	61	62	63	64	37	2
28	29	30	31	32	33	34	35	36	1

19	18	17	16	15	14	13	12	11	10
20	51	50	49	48	47	46	45	44	9
21	52	75	74	73	72	71	70	43	8

Shading in nines

22	53	76	91	90	89	88	69	42	7
23	54	77	92	99	98	87	68	41	6
24	55	78	93	100	97	86	67	40	5
25	56	79	94	95	96	85	66	39	4
26	57	80	81	82	83	84	65	38	3
27	58	59	60	61	62	63	64	37	2
28	29	30	31	32	33	34	35	36	1

1. 5th line down

2. 6 and 7th lines down

P30 Complete the rest answers are: 36, 49, 64, 81, 100, 121, 144, 169, 196, 225

P31 64, 125, 216, 1000

P32/3 This is called The Sieve of Erastosthenes. All of the remaining numbers are prime numbers and have 2 factors only and those factors are different.

P34 365 years

P35 10, 15, 21, they are called triangle numbers.

P36

1.

1	1	2	**3**	5	8

2.

1	2	**3**	**5**	8	**13**

3.

2	**3**	5	8	**13**	**21**

4.

8	**13**	**21**	**34**	**55**	**89**

P37	**1.**	13	15	17	odd numbers
	2.	12	14	16	even numbers
	3.	40	45	50	multiples of 5
	4.	49	56	63	multiples of 7
	5.	72	81	90	multiples of 9
	6.	96	108	120	multiples of 12
	7.	90	105	120	multiples of 15
	8.	96	112	128	multiples of 16
	9.	78	91	104	multiples of 13

10.   48   56   64   multiples of 8

11.   9   10   11   add 1

12.   21   24   27   multiples of 3

13.   9   5   1   subtract 4

14.   21   14   7   subtract 7

15.   32   29   27   subtract 2

16.   44   35   26   subtract 9

17.   25   16   9   square numbers

18.   51   61   71   add 10

19.   666 6660   6 666 6600   666 666 000   multiply by 10

20.   0   20   0   steps in increments of 5 i.e. +5, -5, +10, -10 etc.

21.   18, 36   multiples of 6

22.   400, 250   subtract 50

23.-30. student's own answers

P41   More Sequences

| 165 | 150 | 135 | 120 | 105 | 90 | 75 | 60 | 45 | 30 | 15 |

| 13 | 22 | 31 | 40 | 49 | 58 | 67 | 76 | 85 | 94 | 103 |

| 12 | 19 | 26 | 33 | 40 | 47 | 54 | 61 | 68 | 75 | 82 |

| 76 | 68 | 60 | 52 | 44 | 36 | 28 | 20 | 12 | 4 | -4 |

| 2.5 | 4 | 5.5 | 7 | 8.5 | 10 | 11.5 | 13 | 14.5 | 16 | 17.5 |

| 7 | 14 | 21 | 28 | 35 | 42 | 49 | 56 | 63 | 70 | 77 |

| 132 | 120 | 108 | 96 | 84 | 72 | 60 | 48 | 36 | 24 | 12 |

| 0 | 2 | 4 | 6 | 8 | 10 | 12 | 14 | 16 | 18 | 20 |

| 10 | 20 | 30 | 40 | 50 | 60 | 70 | 80 | 90 | 100 | 120 |

Number sequences - counting in 2s

| 2 | 4 | 6 | 8 | 10 | 12 | 14 | 16 | 18 | 20 | 22 |

| 3 | 5 | 7 | 9 | 11 | 13 | 15 | 17 | 19 | 21 | 23 |

| 32 | 34 | 36 | 38 | 40 | 42 | 44 | 46 | 48 | 50 | 52 |

| 31 | 33 | 35 | 37 | 39 | 41 | 43 | 45 | 47 | 49 | 51 |

Number sequences - counting in 3s

| 3 | 6 | 9 | 12 | 15 | 18 | 21 | 24 | 27 | 30 | 33 |

| 5 | 8 | 11 | 14 | 17 | 20 | 23 | 26 | 29 | 32 | 35 |

| 11 | 14 | 17 | 20 | 23 | 26 | 29 | 32 | 35 | 38 | 41 |

| 41 | 44 | 47 | 50 | 53 | 56 | 59 | 62 | 65 | 68 | 71 |

| 0 | 6 | 12 | 18 | 24 | 30 | 36 | 42 | 48 | 56 | 60 |

| 7 | 14 | 21 | 28 | 35 | 42 | 49 | 56 | 63 | 70 | 77 |

| 14 | 28 | 42 | 56 | 70 | 84 | 98 | 112 | 126 | 140 | 154 |

| 10 | 17 | 24 | 31 | 38 | 45 | 52 | 59 | 66 | 73 | 80 |

| 14 | 19 | 24 | 29 | 34 | 39 | 44 | 49 | 54 | 59 | 65 |

| 36 | 34 | 32 | 30 | 28 | 26 | 24 | 22 | 20 | 18 | 16 |

| -4 | -7 | -10 | -13 | -16 | -19 | -22 | -25 | -28 | -31 | -34 |

| 10 | 13 | 16 | 19 | 21 | 24 | 27 | 30 | 33 | 37 | 40 |

| 38 | 42 | 46 | 50 | 54 | 58 | 62 | 66 | 70 | 74 | 78 |

| -94 | -90 | -86 | -82 | -78 | -74 | -70 | -66 | -62 | -58 | -54 |

| 23 | 28 | 33 | 38 | 43 | 48 | 53 | 58 | 63 | 68 | 73 |

| 12 | 16 | 20 | 24 | 28 | 32 | 36 | 40 | 44 | 48 | 52 |

| 87 | 80 | 73 | 66 | 59 | 52 | 45 | 38 | 31 | 24 | 17 |

| 93 | 85 | 77 | 69 | 61 | 53 | 45 | 37 | 29 | 21 | 13 |

| 41.5 | 42 | 42.5 | 43 | 43.5 | 44 | 44.5 | 45 | 45.5 | 46 | 46.5 |

| 74 | 68 | 62 | 56 | 50 | 44 | 38 | 32 | 26 | 20 | 14 |

| 100 | 150.5 | 201 | 251.5 | 302 | 352.5 | 403 | 453.5 | 504 | 554.5 | 605 |

| 12.5 | 25 | 37.5 | 50 | 62.5 | 75 | 87.5 | 100 | 112.5 | 125 | 137.5 |

| 10 | 22.5 | 35 | 47.5 | 60 | 72.5 | 85 | 97.5 | 110 | 122.5 | 135 |

| 128 | 150 | 174 | 194 | 216 | 238 | 260 | 282 | 304 | 326 | 348 |

A crossword-style word grid containing:

- **ADD**
- **PERCENT**
- **MULTIPLICATION**
- **FACTOR**
- **DIVIDE**
- **SPREADSHEET**
- **SUBTRACT**
- **MULTIPLE**
- **PERCENTAGES**

Number Patterns and Sequences

Across

1 one hundred and fifty five thousand five hundred and fifty five in numbers
2 5001 + 554
3 2 x 2 then add 1
4 1000 + 101
5 110 -9
7 1000 + 250
8 1/2 a million in figures
9 one thousand nine hundred and ninety nine in figures
10 four thousand five hundred in figures
12 66 x 10
13 5 x 12
16 220 -18
17 sixty five thousand in figures
18 5 x 1000
20 666-222
22 4 x 11
23 20 000 + 5552
24 fifteen thousand in figures
25 33 x 20
26 2900 -1 6
27 one hundred and fifty thousand in figures

Down

1 twelve thousand and one + four hundred and ninety nine
2 five thousand + one hundred and one
3 fifty one thousand and forty six in figures
4 1600 - 90
5 one million and ninety thousand five hundred in figures
6 2000 - 10
7 one thousand and fifty - 4 in figures
8 50 x 10
10 23 x 2
11 5 x 10
12 six million five hundred and sixty thousand four hundred and twenty in figures
14 50 x 10
15 two million two hundred and thirty four thousand five hundred and sixty eight in figures
16 2 x 11
19 99 999 + 1
20 5000 - 432
21 4500 + 64
22 1999+1
23 111 x 2
25 (5 x 12) + 8